YOUR KNOWLEDGE HAS VALUE

- We will publish your bachelor's and master's thesis, essays and papers

- Your own eBook and book - sold worldwide in all relevant shops

- Earn money with each sale

Upload your text at www.GRIN.com
and publish for free

Bibliographic information published by the German National Library:

The German National Library lists this publication in the National Bibliography; detailed bibliographic data are available on the Internet at http://dnb.dnb.de .

Imprint:

Copyright © 2016 GRIN Verlag, Open Publishing GmbH
Print and binding: Books on Demand GmbH, Norderstedt Germany
ISBN: 9783668214224

This book at GRIN:

http://www.grin.com/en/e-book/321326/you-reap-what-you-sow-reasons-for-low-profit-margins-among-smallholding

M. Mizanur Rahman, Tasfi Sal-sabil

You reap what you sow? Reasons for low profit margins among smallholding rice farmers in the Philippines

A case study of farmers in the municipality of Calumpit

GRIN Publishing

Reasons for receiving low profit margins by the smallholding farmers though the farmgate price of rice is high: A case study of farmers in the Municipality of Calumpit of Philippines

M Mizanur Rahman[1]

and

Tasfi Sal-sabil[2]

[1]Post-graduate student of NOHA Int. Masters in Humanitarian Action at the Ruhr University of Bochum, Germany

and

[2]Masters in Development Studies at the University of Dhaka, Bangladesh

Using a case study approach, this article explores why the smallholding rice farmers in the Philippines receive low profit which is even as less as the minimum wage when the retail price of rice is significantly higher in comparison to those of nearby countries. In the Calumpit municipality of Bulacan province of the Philippines, the study uses mixed design with both qualitative and quantitative tools. It is a general perception that presence of middlemen in the rice value chain minimizes the profitability of the farmers but in Philippines, the scenario is not the same. This study explains how the limited or almost no access to the factors of production can make the smallholding farmers dependant on the other actors who eventually extracts the major profit share from the value chain. It also emphasizes on the promotion of multi-cropping by making water and other inputs available for farming, which eventually can make smallholding farming profitable in the study area.

Key words: Agriculture, smallholding farmers, rice farmers, poverty, profit, middlemen

INTRODUCTION

Why the country where the International Rice Research Institute (IRRI) was established in 1959, is now a net importer of rice is a paradox. Rice is a popular diet in Philippines and apart from being a food staple, it is also an important economic commodity. Smallholding farms of this country not only meet the food demand of the country but also create employment opportunities for approximately 80 percent of rural people. Although in the 70s, with the spread of the green revolution, the country had experienced a significant growth in agricultural production and attained food sufficiency; in recent years it has turned into a rice importing country.

Smallholders often have limited access to markets for both inputs and outputs, and this has a significant effect on their production activities (Fan et al., 2013). On the other hand, they suffer from high production cost. In developing countries, increasing input costs for fertilizers, pesticides, price of electricity and diesel have frustrated farmers. As a result, many farmers are being pushed either toward cash crop production (which does not contribute to national food security) or out of agriculture altogether. Rahman et al. (2012) analyze the case of Bangladeshi small rice farmers and note that in the face of rising production costs and low returns from production, retaining smallholding farmers in rice production is a challenge in many developing countries. In Philippines, both the farmgate price and retail price of rice is higher than that of the neighboring countries like Vietnam and Thailand but still the small farmers are not happy with the profit they gain from rice production here.

However, the current major rice exporting countries are those which are located within the Asian landmass, largely irrigated by rivers. These countries are mainly Thailand, Vietnam, Cambodia, and Myanmar, those which are traversed by the mighty Mekong River. Many of the land areas of these countries are irrigated naturally i.e. rain fed and irrigated from river throughout the year and so they enjoy some comparative advantages for rice production and have lower production cost in comparison to many other countries.

A few years ago, the Philippines was also one of the rice exporting countries but now it is a net rice importing one. There are several reasons for this; the first is agricultural price policies. Agricultural price policies have a significant role in determining rice price and also rice production of a country. As a state, the Philippines has not been successful in controlling or influencing rice production or in setting (fixing) rice prices. Although, between the 1940s and 1970s there was a new rice production regime in terms of boost in production spurred by land reforms and the green revolution, the National Food Authority (NFA) could not regulate rice price in recent years (as it did as recently as between 1990 to 2009). In recent years, the NFA had been able to procure less than 2 per cent of the total rice production of the country (Tolentino and Pena, 2011). Governments need to maintain a balance between the interest of the poor rice producers and the poor non-producers, who seek higher rice prices, and those who are largely the landless and urban poor consumers, who need low rice prices. This is the reason why the NFA adopted the policy to 'buy high and sell low', thereby benefiting both producers and consumers. It bought rice from farmers with a higher price and sold to poor consumers at low prices via rice subsidies. But due to having less capacity, the volume of procurement of NFA was very small in comparison to net production. It could not really have a substantial impact on the market and eventually, the rise in the domestic production of rice lowered the rice prices and farmers were dis-incentivized.

In recent years the growth rate of rice production in the Philippines has been slower in comparison with that of population. Moreover, the demand for rice has increased as the income level of people has risen and it creates extra demand for different types of foods that are processed from rice. In 1995, when Philippines acceded to the World Trade Organization (WTO) and promised to remove all quantitative restrictions and reduce tariff protection, scope was created to import rice to meet the increased level of demand and this is how, the country moved to be rice importing one though, both yield and cultivation area have also risen over the years resulting to a higher production. From 2004 to 2014, the cultivated area rose from 4100 ha to 4692 ha as the IRRI (2014) data indicates.

Despite the series of institutional rearrangements, the achievement of Philippines in terms of food security is disappointing. From 1998 to 2009, the percentage of people suffering from hunger has risen from 10% to 24.2% (Tolentino and Pena, 2011). This raises questions about why despite the government's declarations to eliminate hunger; it gradually worsened in this country. A key problem is the price of rice, which is significantly higher in Philippines than that of other countries.

In this paper we examine the reasons for the disincentives served by changes in rice price policy and squeeze on the profits of farmers through high costs of production and un-decent value chain relations between traders and farmers. We attempt to explain why smallholding rice farmers receive low margin using the case study of the Municipality of Calumpit in the Philippines.

METHODOLOGY

Study area

Based on the concentration of rice farming, the study has selected five barangayas i.e. Gugo, Longos, San Jose, San Miguel and Santa Lucia of Calumpit Municipality of Bulacan (as shown in Figure 1). The municipality of Calumpit is strategically located with three exit and entry points from North Luzon Expressway, Tabang, Sta Rita, Pulitan Exit. This is generally characterized by flat ground surface with the whole municipality having slopes from 0 to 3 percent. Available data indicates that out of the total 5,625 hectares, 5,434.31 hectares or 96.61 percent land area fall under 0-20 meters evolution. This is because of the town's

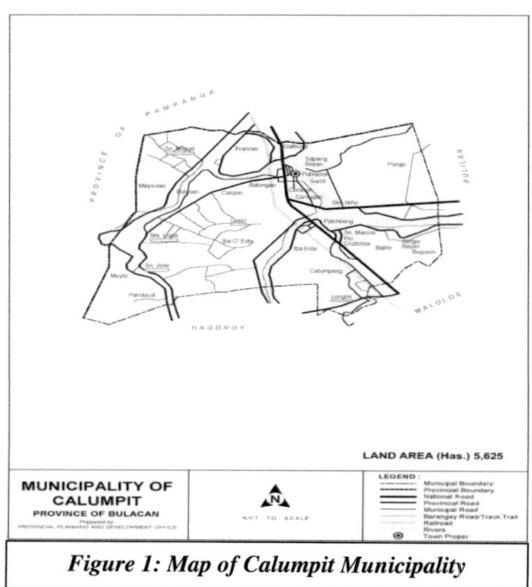

Figure 1: Map of Calumpit Municipality

Source: http://www.calumpit.gov.ph/images/
barangay/administrative%20map_1.jpg

priority of the Manila Bay Area. And thus, this plain land is a fertile and suitable land for rice cultivation. (Municipality of Calumpit, 2014).

The soil of Calumpit is classified as Bigas Clay loam consisting of 184513 hectares of land area of which 3.27 percent, Quingua silt loam with 3790.72 hectares or 67.39 of total area. The Quingua loam has solum depth of over 50 cm and has a loose silt loams topsoil and a slightly composed of heavy silt subsoil and a slightly compact clay loam. This soil has low organic matter content and moderate natural fertility. The available plea porous content is adequate for both upland crops and especially rice (Municipality of Calumpit, 2014).

At the Municipality of Calumpit, maximum land area of 67.66% is used for agricultural production while residential area,rivers and creeks cover more than 10% each. Land area and population of the five selected barangayas area is given in Table 1.

Table 1: Statistics of population and land area

SL	Name	Land area (Ha)	Population
1	Gugo	204.5	1,731
2	Longos	152.6	3,301
3	San Jose	314.6	5,480
4	San Miguel	296.2	5,670
5	Santa Lucia	182.1	2,580

Source: The Municipality of Calumpit, 2010

Sample size and tools of the study

The study adopted a mixed design with the mixture of quantitative and qualitative tools. For the survey, stratified random sampling was used and it gathered specific and detail information on production costs, average farmgate price etc. The list of farmers in these villages was collected from Municipality Agricultural Office (MAO). For selecting respondents we followed two inclusion criteria- research participants must be a small holding rice farmers and for whom rice farming is the primary income source. Based on these two criteria, we found out that 115 farmers meet both of our requirements. The farmers who did not meet both or any of the requirements were excluded. Finally, out of those 115 farmers, 50 have been selected randomly from five villages 10 from each.

Apart from this, five barangay captains were interviewed with a semi-structured questionnaire and five Focus Group Discussions (FGDs) were conducted in all the selected villages. Participants of the FGDs were randomly picked among the villagers. All the participants of the FGDs were either rice farmers or having good knowledge about the current situation of rice farming at that area. In the interviews, the questions were set to understand the detail of the situation under which the smallholding farmers are obliged to sell their productions to some particular people and also to understand the relationship among the actors in the value chain of rice at the study area. FGDs were helpful for having a better understanding of the agricultural market and current challenges facing farmers in regards to rice farming and marketing. FGDs were also used to get the data for value chain analysis and seasonal calendar. The total number of the FGD participants was 64 where 56% were female and 44% were male.

Scenario analysis of rice production in Philippines

The country produced 10.64 million tons of milled rice from 4.56 million hectares in 2011, based on an average yield of 2.33 tons per hectare. In the last decade, the total harvested area in the Philippines has increased steadily (Figure 2). Likewise, the total rice production grows at 2.08% annually, 1.73% of which comes from yield improvement and 0.34% from an increase in area harvested. Total rice consumption (12.39 million tons in 2011) grows at 1.54% per year due solely to population growth, as per capita consumption declines by 0.23% annually. Despite projected gains in rice output, the country is expected to remain a rice importer over the baseline period as the population continues to grow and a safe level of stocks is maintained for food security (Wailes and Chevez, 2012)

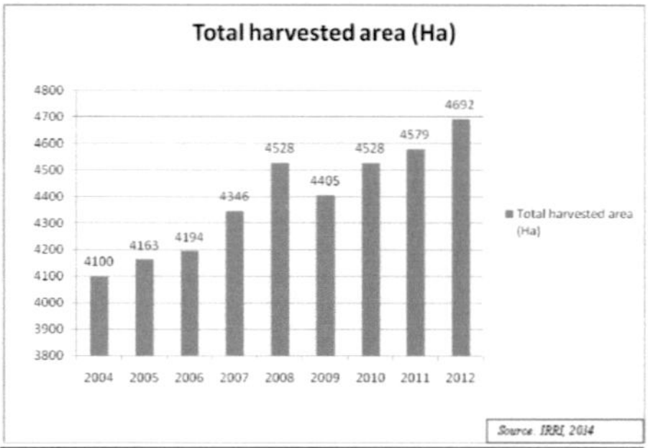

In the same way the total rice production of the country has increased steadily. According to IRRI (2014), rice production of the country has increased from 14,500 tons to 18,016 from 2004 to 2012. Only in 2009, this steady growth was hampered as the Philippines suffered from an El Niño-induced drought, drying up watercourses and irrigation systems in some of the most productive rice areas in Luzon (Redfern *et al*, 2012)

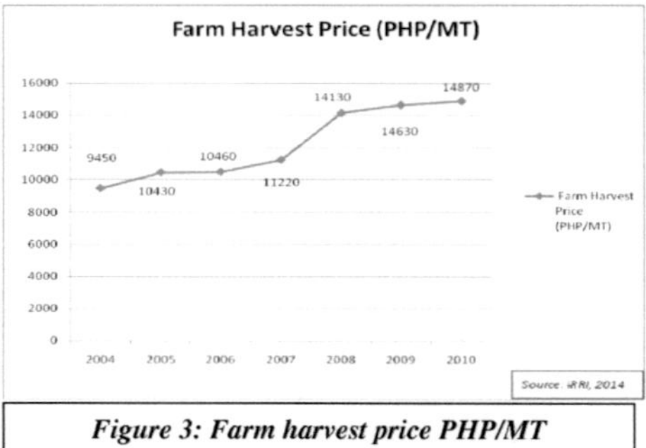

Figure 2: Total harvested area (Ha)

On the other hand, the farm harvest price has also increased over the years. According to the IRRI (2014), the farm harvest price of rice has increased by 8% in every year in an average from 2004 to 2012 (Figure 3). But it is important to note that during this time, the economy of Philippines experience an inflation rate 5.03% in an average. So, it can be said that, the income of the farmers did not increase significantly during the last decade.

Figure 3: Farm harvest price PHP/MT

Value chain analysis of rice at Calumpit municipality

There are broadly six main stakeholders in the value chain of rice as shown in the Figure 5. After farmers harvest their rice, small traders buy it and sell it on to the big traders who are often involved with rice processing. This year, the non-processed rice price was sold from 22 to 24 Philippino Peso (PHP) per kilogram (kg). If the rice quality is very good, small traders pay them 24 PHP when lower quality rice gets 22 PHP per kilogram. In the field it has been observed that in many cases, the small traders fix the price and they try to establish that the rice quality is not so good that they can pay 24 PHP/kg and thus they tried to buy from the farmers with as low a price as possible. However, these middle men have very small profit margin because in this year, they had to sell their rice to the millers/processor or the big traders at 24 PHP/kg as well. So, where is the profit margin for these small traders? They get a certain amount of commission which is 20 cents per kg from the traders whom they sell the rice to. That is why they try to lower their purchasing price in order to have a greater margin. If they can really establish that the rice quality is poor and therefore they can pay only 22 PHP to the farmers and on the other hand they can sell it to the millers with 24 PHP, they can have additional 2 PHP profit per kilogram. So, in total for one kilogram, they can get 2.2 PHP as profit and if they buy it with 23 PHP/kg, the profit margin becomes 1.2 PHP/kg (Detail can be seen in Table 2).

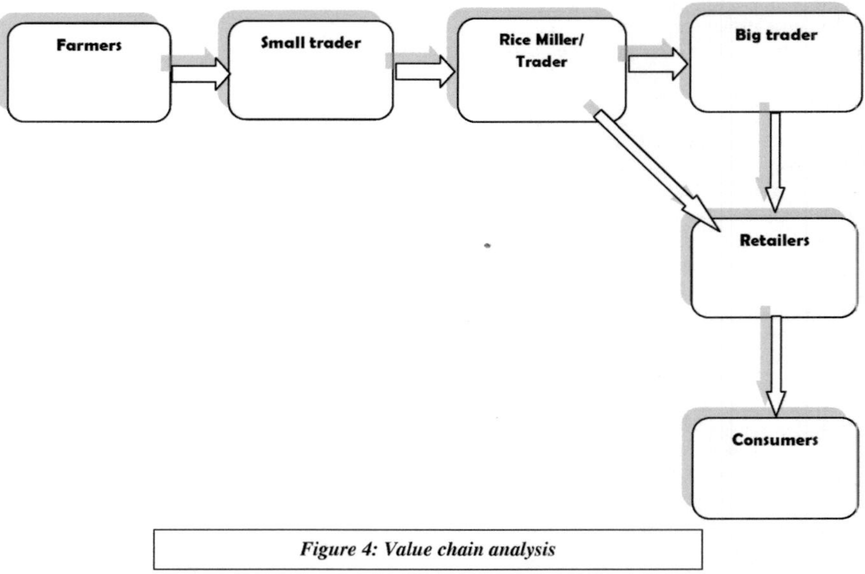

Figure 4: Value chain analysis

To many of the farmers, the small traders profit is rational, because the traders need to make a profit too and need to cover the cost of their own capital. After the millers or the processors get the rice from the small traders, they (big traders) process them and then some of them directly give the processed rice to the retailers and some supply to the other big traders who later on sell rice to the retailers.

Table 2: Transaction and associated price

SL	Transaction between	Cost to the seller/KG (including selling price and associated cost)	Price seller receives/KG
1	Farmers and small traders	20	22.5
2	Small traders to millers/processor	23	24.20
3	Millers to big traders	41.5	43.5
4	Big traders to retailers	44	45
5	Retailers to consumer	45.20	46

Note: This calculation is based on 2014 price and only the average price

If the millers buy the good quality of rice with 24 PHP/kg, and they pay 20 cents as commission to the small traders, the net cost of purchasing one kg of rice for the millers becomes 24.20 PHP/kg. So, for purchasing one sack (50 kg) of rice, a miller spends 1210 PHP. After processing, they get 30 kg of processed rice out of one sack i.e. 50 kg or rice. For processing and other costs, they pay in an average 35 PHP for one sack of rice. So, the total investment for one sack of rice becomes 1245, which is the production cost of 30 kg of processed rice. It means that for the miller, the cost for one kg of rice is 41.5 PHP but his selling price to the big traders is only 43.5 to 44 PHP with a profit margin of 2 to 2.5 PHP per kg. Some of these millers or processors directly sell to the retailers with 44 to 44.5 PHP rate but they prefer to sell to the big traders who then, accumulate and distribute to different region and market retailers with 44.5 to 45 PHP and finally the retailers sell to the consumer with 46 PHP. So, in the entire cycle, no-one is extracting very unusual profit and there is hardly any scope for any of the agents in the chain.

High cost due to extractive tied arrangements

Most of the farmers at the Calumpit municipality are smallholders and often cultivate rented in land for which they need to pay a high rent. Many of the farmers are dependent on others for credit otherwise; they would not been able to afford the land rental and other production costs. Very few farmers borrow from local money lenders or the bank; instead they borrow through various arrangements. For example, the landless rent in land on credit on the condition that they will pay a certain (10 sacks for one hectare in most cases) amount of rice after harvesting. They also can get fertilizer on credit. Because the owner of harvesting and threshing machines extend credit for production including fertilizer, the landless farmers are forced to spend a large amount of money to hire in the machines with a further 20 sacks of rice out of 100 sacks. So, when for a landless farmer the owner of the land and machine and also creditor is the same person, the farmers need to pay 40 sacks of rice out of produced 100 sacks to the same person. Thus in this entire deal, the winners are the landowners and machine owners who extend the credit.

For one hectare land cultivation for rice, the farmers need to spend almost 54,000 PHP in production costs alone. If they need to rent in land, they need to spend an additional 12000 PHP or 20 sacks of rice in every 100 sacks. In 2014, the average production of one hectare was 100 sacks i.e. 5000 kg of rice out of which, they had to pay at least to the harvester and thresher 20 sacks. So, at the end, they had 80 sacks i.e. 4000 kg of rice the price of which was 92000 PHP (at an average rate of 23 PHP/kg). Thus, the profit for cultivating one hectare of rice becomes 38000 PHP (the profit will be lower around 26000 PHP if farmers need to rent in land) if there are no other expenditures like the interest for the loans. Maximum of the farmers have 1.5 hectare of land area which can give him 57000 PHP profit in a year at best. In other words, he or she gets 4750 PHP in a month to run family, which is lower than the minimum wage[1]. It creates disincentive for the farmers as with all their effort, they cannot even earn the minimum wage and it is not even sufficient to meet the food demand of the family members. Thus, this profession cannot even decrease hunger among the small rice farming people in the Philippines.

RESULTS AND DISCUSSION

Dependence of farmers on factors of production

In the study area, 83 per cent of the respondents have been seen entering into contractual arrangements with local elites who provide credit to farmers in kind –providing fertilizers and pesticides to the farmers on the condition that they (farmers) will pay them with rice once they get their harvest. Apart from this, the most severe one is helping the farmers by providing the service of harvesting and threshing from the part of those lenders. Usually, these contracts are exploitative and work in favor of the resource rich lenders. But still, according to Eswaran and Kotwal (1985), farmers continue with these types of contracts for three reasons – (1) tradeoff between risk sharing and transaction costs, (2) screening of workers of different qualities and (3) market imperfections for inputs besides land.

Particularly in Gugo, the same people do all these things i.e. they provide credit, fertilizers and pesticides and they also provide support for harvesting and threshing rice. This is how, from one hectare of land, they manage to get 20 sacks of rice when the farmers get 80 sacks for meeting the production cost and also for securing the whole year family expenses. However, agricultural price policies also have a significant role in determining rice price and also rice production of a country.

In principle, governments need to maintain a balance between the interest of the poor rice producers and the poor non-producers who are basically urban poor consumers. But as the amount of procurement at national level was very small, due to the low capacity of the NFA, the poor farmers in the study area did not get the benefit of it. So, the macro level policy issues found it difficult to reach the grassroot level especially when food policy of the Philippines is concerned. NFA adopted policy to 'buy high and sell low' could not really have an impact on the market and life of the small rice farmers of the Calumpit Municipality. That is why, people have been found pessimistic of the government support as one respondent noted, 'government is only for the rich, not for us', which questions governance issues of the country.

This can lead to a further discussion on the NFA functioning in the country. According to Cororaton, (2004),

---

[1] Minimum wage in Philippines is 466 PHP/day. For detail: http://www.gmanetwork.com/news/story/342856/news/nation/ncr-minimum-wage-now-at-p466-day-after-p15-increase

'The government policy on rice is relatively more successful in defending consumer price ceilings than price floors. As a result, farm prices remained below rice support prices because of a number of reasons, which include inadequate NFA procurement budget, delayed timing of NFA purchases, etc. Thus, margins are squeezed, resulting in reduced investment in postharvest facilities and less planting because of unattractive price to farmers. On the other hand, in the long-run the consumer oriented pricing policy does not benefit consumers because it reduces rice availability'.

During the interviews, some local experts argued that they see a minimum level of government influence in the local rice market of Bulacan. They noted that as a state, the Philippines has not been that successful in controlling or influencing rice production or even rice price fixation at the national level. The local government has been found to undertake some agricultural promotional activities, but they lack any intention to influence the local agricultural market system. The local government authorities are aware of the facts that how the small holding rice farmers are in many ways dependent on that particular group of people but without any further guidance or instruction from the provincial or central government, they do find anything to deal with these issues. So, some sort of strategic position on how to support the small holders to recover their dependence on others for those factors of production of the government and guidance towards the local government as per that from the part of the higher level of authority would have worked well.

Soil condition, irrigation crisis and cropping intensity

Geographically Philippines is highly vulnerable to multiple hazards as it lies on the western rim of the Pacific and along the circum-Pacific seismic belt, subjecting it to typhoons, earthquakes, floods, volcanic eruptions, droughts, and other natural hazards. Dilley (2005) addressed in the 'Natural Disaster Hotspots: a Global Risk Analysis' book that Philippines is a highly risk country due to flood and cyclone. In fact, according to World Bank's Natural Disaster Hotspot list (Dilley, 2005), Philippines is ranked eighth due to its most exposed to multiple hazards, with 268 recorded disasters over the last three decades. Moreover, maximum of the land area of Calumpit is low lying which adds some more vulnerability to the agricultural land of this area. Being low lying, this place is highly vulnerable to water stagnation during the rainy season. There is only one river (Calumpit River) on which the irrigation of this area is dependent. The longest river system in Bulacan, Calumpit River traverses the towns of Calumpit, Pulilan and Plaridel on the east, Paombong and Hagonoy in the West and winds up through Apalit, Macabebe and Masantol, Pampanga. Therefore, people cannot go for multi-cropping in the Calumpit area. They get only one crop of rice in a year which covers November to February and the rest eight months their land remains without cultivation and maximum time under water. So, people here remain almost unemployed for eight months in a year though during the rainy season, they can go for fishing but it is only subsistence and does not give any good amount to live on.

Water shortage and mismanagement are becoming major constraints in the country for low crop yields (Qureshi, et al., 2012). Though there was one project of the government i.e. NIA[2], which was to ensure irrigation for all the farmers but it has been able to cover only one third of the

---

[2] The National Irrigation Administration, detail can be found at: http://www.nia.gov.ph/

farmers. Some of the respondents claim that this project could not meet the entire demand of the area due to less amount of allocated fund for this by the government.

Natural Calamities

In the recent years, rice production systems of the South-east region have become increasingly threatened by the effects of climate change (Masutomi et al., 2009). Natural disasters, as consequences of climate change, are common phenomenon in the Philippines. This land of agriculture has a long historical relationship with natural calamities. Sivakumar (2005) explained that natural disasters cause environmental degradation which in turn contributes to the disaster vulnerability of agriculture, forestry and rangelands. The similar picture is also observed in the study area.

The climate of Calumpit is similar with that of the rest of the other municipalities in the province of Bulacan. It is characterized by two (2) distinct seasons namely; the rainy and the dry. The rainy season starts from late May and ends around November, the dry season is from December to April (Detail can be seen in Table 3). The average annual rainfall is 255.3 millimeters (10.05 in) with the month of August having the highest month average rainfall is about 304 millimeters (12.0 in). The annual number of rainy days is 175 days.

Table 3: Seasonal calendar

Nurturing of paddy	January
Harvesting	February
No irrigation water except salty sea water from the sea	March
No irrigation water except salty sea water from the sea	April
No irrigation water except salty sea water from the sea	May
No irrigation water except salty sea water from the sea	June
Flood due to heavy rain	July
Flood due to heavy rain	August
Flood due to heavy rain	September
Land preparation for rice cultivation	October
Planting of paddy	November
Nurturing of paddy	December

Rice farming is the most common agricultural practice of the people in this municipality though it can occupy only four months of them and the rest of the time they remain without any regular income generating works. The seasonal calendar below shows that in the month of October, the farmers start preparing their land, in November they plant their paddy, rest two months they nurture the paddy with proper fertilizer, pesticides and so on. Finally, in February they harvest their rice.

Credit unavailability and dependence on local money lenders

Inaccessibility to credit from formal lending institutions is one of the major problems of the small farmers in the developing countries (Amjad & Hasnu, 2007; Atieno, 2001; Braverman & Guasch, 1986; Yaron & Mundial, 1992) and in the study area as well. As 91 per cent of the smallholders in the study area do not have collateral, they cannot take loans from the banks or from the Micro Finance Institutions (MFI). Commercial banks and other formal institutions fail to cater for the credit needs of smallholders, however, mainly due to their lending terms and conditions and it is generally the rules and regulations of the formal financial institutions that have created the myth that the poor are not bankable, and since they can't afford the required collateral, they are considered uncreditworthy (Atieno, 2001). Evidences suggest that large farmers have better access to formal sources, on the other hand, majority of the small farmers not only have limited access to formal sources but their access to informal sources i.e. friends, relative and landlords, is also highly restricted (Amjad & Hasnu, 2007). In the study area, 42 per cent of the farmers have reported to borrow money from friends and relatives for meeting the agricultural expenses when 28 per cent of them go to the local money lenders to borrow money with 20% interest monthly. People do not have an idea about how high this interest rate is. They just get this loan as this is very easy to get and there is no procedural problems like the bank or other MFIs.

However, the above discussion implies as observed by Bautista and Javier (2005), that over the last few decades, rice production practices in the Philippines have changed mainly due to introduction of technologies and some government programs envisioned responding to the dynamic challenges and needs of the Filipinos and has directly affected the production cost. The small holding farmers have been affected significantly due to their unavailability of credit and more the agricultural practice has adopted technology, it always created some inconvenience for the farmers.

Conclusion

Increased level of price does not necessarily bring good for the smallholders. Recent researches in Bangladesh and Malawi show that an increased level of staple crop price results in a higher welfare loss for small landholders (Karfakis, et al., 2011). Similarly in Philippines, higher level of rice price has not been able to supply minimum level of wage to the farmers. When on the other hand, evidence from Ghana shows that higher maize prices have the largest adverse welfare effects on urban, female-headed, poor, and small farm households because these groups are traditionally net buyers of maize (Minot & Dewina, 2013). It indicates that the poor net consumers of staple food are the highest victim of staple price hike. So, with increased level of price neither the poor producers nor the poor consumers are benefited.

Perdana et al. (2012) emphasize on the bargaining power of the farmers and note that farmers must deal with the overwhelming profit-eroding power of buyers and the intermediaries. But it is not relevant for the smallholding farmers at the study area. Huge cost of production works as the main adverse force for the smallholding farmers in Philippines. However, this cost is also very closely associated with their dependence on others for the factors of production. So, while finding out strategies of improving the profitability of smallholding rice farmers in Philippines, cutting their cost of production by making them self-reliant for factors of production needs to be given priority among the other possible options.

However, the respondents came up with two potential solutions by themselves when one can be solved internally and for the other, government or external intervention is essential. If they are ensured with the easy access to those factors of production with a minimum cost, their production cost will be significantly diminished and thus with the existing farmgate price, they can gain higher level of profit. Apart from this, increasing cropping intensity by making irrigation available can also be a way for making farming profitable. Analyzing the case of India and the Philippines, Bouman and Toung (2001) emphasized on water storage and opined that total rice production can be increased by using water saved in one location to irrigate new land in another. This study also suggests the similar or undertaking a large scale irrigation project. However, further study needs to be undertaken for exploring these two potential and possible scopes in Philippines.

Acknowledgements

The authors express their gratitude to the participants of the research, Professor Ron Chua of Asian Institute of Management, Lucia de Corta from University of Bath, Municipality Agricultural Office (MAO) of Calumpit and the Stephen Zuellig Graduate School of Development Management (SZGSDM) of Asian Institute of Management, Philippines.

REFERENCES

Amjad, S., & Hasnu, S. A. F. (2007). Smallholders access to rural credit: Evidence from Pakistan. The Lahore Journal of Economics, 12(2), 1-25.

Atieno, R. (2001). Formal and informal institutions' lending policies and access to credit by small-scale enterprises in Kenya: An empirical assessment (Vol. 111): African Economic Research Consortium Nairobi.

Bautista EU, Javier EF (2005). The Evolution of Rice Production Practices (No. DP 2005-14). Philippine Institute for Development Studies.

Bouman BAM, Tuong TP (2001). Field water management to save water and increase its productivity in irrigated lowland rice. Agricultural Water Management, 49(1):11-30.Crossref

Braverman, A., & Guasch, J. L. (1986). Rural credit markets and institutions in developing countries: Lessons for policy analysis from practice and modern theory. World Development, 14(10), 1253-1267.

Cororaton CB (2004). Rice reforms and poverty in the Philippines: a CGE analysis: ADB Institute Discussion Paper No. 8

Dilley, M. (2005). *Natural disaster hotspots: a global risk analysis* (Vol. 5): World Bank Publications.

Eswaran M, Kotwal A (1985). A theory of contractual structure in agriculture. The American Econ. Rev., pp.352-367.

Fan S, Brzeska J, Keyzer M, Halsema A (2013). From subsistence to profit: Transforming smallholder farms (Vol. 26). Intl Food Policy Res. Inst.

IRRI. (2014). Resource: World Rice Statistics (Publication. Retrieved 01 July 2014, from IRRI: http://ricestat.irri.org:8080/wrs2/entrypoint.htm)

Karfakis P, Velazco J, Moreno E, Covarrubias K (2011). Impact of increasing prices of agricultural commodities on poverty. ESA-FAO working paper series, pp.11-14.

Masutomi Y, Takahashi K, Harasawa H, Matsuoka Y (2009). Impact assessment of climate change on rice production in Asia in comprehensive consideration of process/parameter uncertainty in general circulation models. Agriculture, Ecosys. Environ., 131(3):281-291.Crossref

Minot N, Dewina R (2013). Impact of food price changes on household welfare in Ghana (Vol. 1245). Intl Food Policy Res Inst.

Municipality Profile. (2010). Bulacan: Municipality of Calumpit

Perdana A, Roshetko JM, Kurniawan I (2012). Forces of competition: smallholding teak producers in Indonesia. International Forestry Review, 14(2):238-248. Crossref

Qureshi AL, Khero ZI, Lashari BK (2012). Optimization of irrigation water management: A case study of secondary canal, Sindh, Pakistan.

Rahman MM, Klees B, Sal-sabil T, Khan NA (2014). Rice, smallholder farms, and climate change in Bangladesh: thala of policy options. J. Indian Res., 2(1):59-66.Crossref

Redfern SK, Azzu N, Binamira JS (2012). Rice in Southeast Asia: facing risks and vulnerabilities to respond to climate change. Building resilience for adaptation to climate change in the agriculture sector, 23, 295.

Simmons P (2002). Overview of smallholder contract farming in developing countries. Rome: FAO.

Sivakumar MVK (2005). Impacts of natural disasters in agriculture, rangeland and forestry: an overview. In Natural disasters and extreme events in Agriculture pp. 1-22. Crossref

Yaron, J., & Mundial, B. (1992). *Rural finance in developing countries* (Vol. 875): Agriculture and Rural Development Department.

YOUR KNOWLEDGE HAS VALUE

- We will publish your bachelor's and master's thesis, essays and papers

- Your own eBook and book - sold worldwide in all relevant shops

- Earn money with each sale

Upload your text at www.GRIN.com and publish for free